EXPERIMENTS UPON MAGNESIA ALBA, QUICKLIME, AND SOME OTHER ALCALINE SUBSTANCES.

Joseph Black, M.D.

Contents

Bibliographic Key Phrases

Magnesia Alba; Quicklime Chemistry; 18th Century Science; Joseph Black; Alembic Club Reprints; Chemical History; Fixed Air; Caustic Alkali; Absorbent Earths; Lime-Water; Elective Attractions; Quantitative Method; Early Chemistry

Publishing Information

(c) 2024 Nimble Books LLC

ISBN: 978-1-60888-347-9

Nimble Books LLC ~ NimbleBooks.com

Humans and models making books richer, more diverse, and more surprising.

Changelog

- Version 0.5.1.0 of Codexes2Gemini moves Most Important Passages and Condensed Matter to the beginning, as they represent the purest distillation of the experience of actually reading the original book.

- Conversation Starters and Request for Feedback sections are added with the appendices.
- And this section, the Changelog, is added to the front matter.

Publisher's Note

In a world increasingly concerned about the environ-ment and sustainability, the question of how to manage and utilize resources efficiently has taken center stage. This book, "Experiments upon Magnesia Alba, Quick-lime, and Some Other Alkaline Substances" by Joseph Black, delves into the very foundations of chemistry and offers invaluable insights into the nature of materials like lime, magnesia, and alkalis. Black's meticulous ex-periments, conducted in the mid-18th century, provide a groundbreaking understanding of the role of fixed air (carbon dioxide) in these substances, paving the way for future advancements in chemical analysis and understanding. Through his innovative quantitative methods, Black unravels the secrets behind the reac-tions and transformations of these materials, revealing how they gain and lose weight, absorb air, and interact with acids. This comprehensive examination of the fundamental properties of lime, magnesia, and alkalis offers a historical lens on the evolution of chemistry and provides a foundation for understanding crucial processes like the production of quicklime, the formation of caustic alkali, and the role of these materials in various chemical reactions. This reprint of Black's seminal work is a must-read for researchers, students, and anyone interested in the history and practice of chemistry, as it presents a compelling narrative of scientific discovery, experimentation, and the profound impact of understanding the nature of everyday substances.

Truth in Publishing (Disclosures)

This is a reprint of a 1756 paper by Joseph Black, a pioneer in the field of chemistry. The paper is full of fascinating observations, and it was historically significant, but, dear reader, be warned: **this is *not* a casual read**.

What's Good:

- **You get a glimpse into the early days of chemistry.** Black's careful experiments and meticulous observations are a window into a time when chemistry was much more of an art than a science.
- **It's a remarkable work for its time.** Black's experiments helped to lay the foundation for our modern understanding of chemical reactions and

the nature of substances like lime and alkalis.

- **It's a treasure for the history of science buff.** For those with a genuine interest in the history of chemistry, Black's paper is a must-read.

What's Less Good:

- **It's a bit of a slog.** Black's writing is formal, dry, and full of long, convoluted sentences. Prepare for a lot of "altho," "thereby," and "wherewith."

- **The paper is filled with outdated terminology and concepts.** For example, Black uses the term "fixed air" to refer to carbon dioxide.
- **It's long and detailed, but sometimes meandering.** Black doesn't shy away from lengthy descriptions of his experiments and observations, and sometimes the details go on and on.

In a Nutshell:

If you're a history of science enthusiast and are prepared for a challenging read, this paper might be for you. But if you're looking for a light, engaging read, this isn't it.

To help with your reading journey, we recommend:

- **Having a strong cup of coffee or tea.**
- **Keeping a dictionary handy.**

- Perhaps indulging in a bit of "Truth in Publishing" before each section.

Most Important Passages

Black's Paper entitled "Experiments upon Magnesia Alba, Quicklime, and some other Alcaline Substances" was read in June 1755, and was first published in "Essays and Observations, Physical and Literary. Read before a Society in Edinburgh, and Published by them," Volume II., Edinburgh, 1756; pp. 157-225. It was subsequently reprinted several times during the life of the author, not only in later editions of these Essays, but also in a separate form. Copies of the original Paper are now very difficult to obtain, and the later reprints have also become scarce.

> The present reprint is a faithful copy of the Paper as it first appeared in 1756, the spelling, &c., of the original having been carefully reproduced.
>
> The Paper constitutes a highly important step in the laying of the foundations of chemistry as an exact science, and furnishes a model of carefully planned experimental investigation, and of clear reasoning upon the results of experiment. It is neither so widely read by the younger chemists nor is it so readily accessible as it ought to be, and the object of the Alembic Club in issuing it as the first volume of a series of Reprints of historically important contributions to Chemistry, is to place it within easy reach of every student of Chemistry and of the History of Chemistry. (Preface)

This is important because it establishes the context of this reprint and its significance to the history of chemistry. It also emphasizes Black's contribution to the development of chemistry as an experimental science.

> In reflecting afterwards upon these experiments, an explication of the nature of lime offered itself, which seemed to account, in

an easy manner, for most of the properties of that substance.

It is sufficiently clear, that the calcarious earths in their native state, and that the alkalis and magnesia in their ordinary condition, contain a large quantity of fixed air, and this air certainly adheres to them with considerable force, since a strong fire is necessary to separate it from magnesia, and the strongest is not sufficient to expel it entirely from fixed alkalis, or take away their power of effervescing with acid salts. (Part II)

This passage marks the start of Black's explanation of the nature of lime. He proposes that the properties of lime and alkalis are explained by their relationship with a "fixed air," which he goes on to demonstrate through careful experimentation.

This supposition was founded upon an observation of the most frequent consequences of combining bodies in chemistry. Commonly when we join two bodies together, their acrimony or attraction for other substances becomes immediately either less perceivable or entirely insensible; altho' it was sufficiently strong and remarkable before their union, and may be rendered evident again by disjoining them. A neutral salt, which is composed of an acid and alkali, does not possess the acrimony of either of its constituent parts. It can easily be separated from water, has little or no effect upon metals, is incapable of being joined to inflammable bodies, and of corroding and dissolving animals and vegetables; so that the attraction both of the acid and alkali for these several substances seems to be suspended till they are again separated from one another. (Part II)

This section is key because it lays out Black's understanding of chemical combination and the role of "acrimony," or attraction, in chemical reactions. He argues that the acrimony of substances is often masked or neutralized when they combine, and can be revealed again by disjoining them. This is a major step in the development of chemical bonding theory.

I resolved however to examine, in a particular manner, such of these consequences as were the most unavoidable, and found the greatest number of them might be reduced to the following propositions:

I. If we only separate a quantity of air when we render them caustic they will be their weight in the operation, but will s of acid as before, and the saturation wil effervescence.

II. If quick-lime be no other than a calca[
its air, and whose attraction for fixed ai]
of alkalis, it follows, that, by adding to
of alkali saturated with air, the lime wil[
its air, and be entirely restored to its o]
condition: and it also follows, that the e[
lime-water by an alkali, is the lime which
water now restored to its original mild an[

III. If it be supposed that slaked lime do[
which are more firey, active or subtile th[
chiefly it communicates its virtues to wat[
uniform compound of lime and water: it fol[
can be dissolved in water, the whole of it
dissolved.

IV. If the acrimony of the caustic alkali [
part of the lime adhering to it, a caustic
ley will
consequently be found to contain no lime, [
lime employed in making it were greater th[
sufficient to extract the whole air of the
of the superfluous quick-lime might possib[
ley as would be dissolved by pure water, o]
as much lime as lime-water does.

> V. We have shewn in the former experiment
> lose their air when they are joined to ar
> separated again from that acid, by means
> air passing from the alkali to the earth,
> acid passes from the earth to the alkali.

This section is a crucial one because it lists out Black's
key hypotheses about the chemical behavior of lime
and alkalis, which he will then demonstrate through
his experimental observations. It establishes the core
arguments he is making about these substances.

> I therefore engaged myself in a set of tri-
> als; the history of which is here subjoined.
> Some new facts are likeways occasionally
> mentioned; and here it will be proper to
> inform the reader, that I have never men-
> tioned any without satisfying myself of their
> truth by experiment, tho' I have sometimes
> taken the liberty to neglect describing the
> experiments when they seemed sufficiently
> obvious. (Part II)

This brief statement is important because it establishes
the methodology that Black uses in his work. He claims
to rely solely on experimentation to confirm his findings,
despite the fact that he may not always describe his
experiments in detail.

> From some of the above experiments, it ap-

pears, that a few alterations may be made in the column of acids in Mr. *Geoffroy's* table of elective attractions, and that a new column may be added to that table, according to the following scheme, where the alkaline substances are all considered as in their pure state and free of fixed air.

Acids.	Fixed air.
Fixed alkali,	Calcarious earth.
Calcarious earth,	Fixed alkali.
Volatile alkali and magnesia.	Magnesia.
	Volatile alkali.
————————-	———— (Part II)

This is a critical section because it marks the culmination of Black's work on the properties of lime and alkalis. He proposes a new table of elective attractions, a crucial part of 18th century chemical theory, which takes into account the role of "fixed air" in chemical reactions. This demonstrates the significant contribution of Black's work to the developing field of chemistry.

Condensed Matter

Joseph Black's "Experiments upon Magnesia Alba,
Quicklime, and some other Alcaline Substances" was
first presented in June 1755 and published in 1756.
The Alembic Club deemed it worthy of reprint in
1898 because the work represents a major step in
establishing chemistry as an exact science. Black's
work exemplifies a carefully planned and detailed
experimental approach, emphasizing quantitative
methodology and drawing conclusions based on
experimental data. In 1898, the work was considered a
model of scientific rigor and crucial to advancing the
field of chemistry. While the original 1756 publication
was difficult to obtain, the Alembic Club's reprint made
Black's pioneering work readily accessible to students
of chemistry and the history of the subject.

Black's experiments on magnesia alba focused on its

reactions with acids and its behavior under heat. He found that magnesia, unlike calcarious earths, dissolves with violent effervescence in acids such as vitriol, nitre, and distilled vinegar, forming unique salts. He concluded that magnesia is different from calcarious earths, and its attraction for acids is distinct. Through experiments involving salt ammoniac and calcarious earths, Black demonstrated that magnesia's attraction for acids is weaker than that of volatile alkali but stronger than that of calcarious earths. He observed that magnesia loses a significant portion of its weight when subjected to intense heat, suggesting the loss of a volatile component. Distillation experiments confirmed that this volatile component is primarily air, causing the calcined magnesia to lose its effervescence with acids. However, by dissolving and precipitating calcined magnesia with an alkali, Black found that it regained its original properties, including effervescence and increased weight, indicating the absorption of air from the alkali. This led him to conclude that alkalis contain fixed air, released when combined with acids.

Black then compared magnesia to other absorbent earths like chalk, earth of animal bones, and the earth of alum. He found that these earths differed in their reactions with acids and their ability to be converted into quicklime. He proposed a theory that lime and alkalis in their ordinary state are saturated with fixed air, and their acrimony is revealed when this air is removed, as in the case of quicklime. He argued that the acrimony of quicklime doesn't arise from any added substance during burning, but is an essential property of the pure earth, becoming evident only after the separation of fixed air. Through experiments involving the saturation of chalk with acids and the subsequent conversion to quicklime, Black confirmed that lime loses air during calcination, but still saturates the same amount of acid. He also demonstrated that quicklime can be restored to its original state by combining it with a fixed alkali, which provides the necessary fixed air.

Black then examined the properties of caustic alkali, finding that it doesn't effervesce with acids and doesn't render lime-water turbid, indicating a lack of fixed air. He also found no evidence of lime in caustic ley, supporting his theory that causticity arises from the removal of fixed air from the alkali. By examining the effects of caustic alkali on epsom salt and chalk solutions, Black further confirmed that caustic alkali separates magnesia and calcarious earths from acids, producing substances free of fixed air. Finally, Black presented a modified table of elective attractions, incorporating fixed air as a factor, illustrating the relative strength of attraction between acids and various alkaline substances. He concluded that the separation of acids from earths in the presence of alkaline substances is likely driven by a balance of forces between the various components, potentially involving both attraction and repulsion.

Joseph Black's revolutionary theory posited that calcarious earths, alkalis, and magnesia, in their natural state, contain a significant amount of "fixed air." This air is strongly bound to these substances, requiring intense heat for its separation, as seen in the calcination of magnesia. Black proposed that the causticity of quicklime, resulting from its calcination, didn't stem from new matter acquired during the process but rather from the exposure of its inherent acrimony following the loss of fixed air. He likened this phenomenon to the loss of acrimony in a neutral salt, where the acid and alkali, individually corrosive, are neutralized upon union.

Black's experiments on the interaction of lime with magnesia and alkalis supported his theory. When magnesia was mixed with limewater, the lime, having a stronger affinity for fixed air, absorbed it from the magnesia, leaving the magnesia devoid of air and rendering the limewater insipid. He observed a similar interaction with alkalis, where quicklime extracted fixed air from alkalis, leading to the alkali's increased causticity.

The experiments conducted by Black to further investigate the implications of his theory were groundbreaking in their meticulousness. He established that caustic lime and alkalis, though devoid of fixed air, retained their capacity to saturate the same amount of acid as their mild counterparts, but without the effervescence that characterized the reaction of mild substances with acids.

Black's work provided a foundation for understanding the nature of alkalis and lime, demonstrating that causticity arises from the absence of fixed air. He also demonstrated the crucial role of fixed air in the formation of salts, explaining why saturation occurs without effervescence in caustic substances. His investigations, marked by the application of quantitative methods, laid the groundwork for a more precise and rigorous approach to chemical inquiry.

Black presented compelling experimental evidence to support his assertion that the causticity in alkalis stemmed from the removal of fixed air. His experiments focused on the behavior of caustic ley and the volatile alkali, demonstrating the impact of fixed air on their properties.

One key observation involved the interaction of *magnesia* with acids. Black noted that crude *magnesia*, containing fixed air, effervesced vigorously when mixed with acids, while calcined *magnesia*, devoid of fixed air, dissolved without any effervescence. This suggested that the effervescence originated from the release of fixed air trapped within the crude *magnesia*.

Furthermore, Black demonstrated that the causticity of alkalis could be induced by removing fixed air. When he mixed quicklime, a calcarious earth devoid of fixed air, with a solution of a fixed alkali, the lime absorbed fixed air, becoming mild, while the alkali lost its fixed air and became caustic. This loss of fixed air resulted in the alkali's enhanced corrosiveness and volatility.

The experiments with borax also supported Black's theory. Borax, a neutral salt containing a fixed alkali and sedative salt, didn't effervesce when mixed with acids, indicating its alkali was devoid of fixed air. When sedative salt was mixed with an alkali saturated with air, a brisk effervescence ensued, further supporting the idea that the alkali, upon losing fixed air, became more reactive.

Black's careful observations and quantitative approach provided strong evidence that the removal of fixed air played a crucial role in the causticity of alkalis. He solidified this concept through the experiments on *magnesia*, caustic ley, and the volatile alkali, leading to a more comprehensive understanding of the properties and reactions of these substances.

Black's experiments aimed to understand the relative attraction of various substances for "fixed air," a term he used for carbon dioxide. He focused on magnesia, lime, and volatile alkali (ammonia). He found that magnesia, like other absorbent earths, combines with "fixed air" and loses it upon calcination, a process that renders it caustic (reactive). When magnesia is combined with an acid and then separated by an alkali, it regains "fixed air" and its original properties.

Black then proposed a modified version of Geoffroy's table of elective attractions, incorporating "fixed air" as a significant factor. This modification places calcarious earth above fixed alkali in the "fixed air" column, reflecting their relative attractions. He also suggests that volatile alkali and magnesia have similar attractions for "fixed air," explaining their behavior in distillation.

Abstracts

TLDR (three words)

Lime is airless.

ELI5

Imagine a special kind of air that's stuck to things like chalk and baking soda. When you heat them up, they lose this air, and they become stronger and more reactive. This is like when you make cookies - the baking soda makes the cookies rise because it loses air and becomes more powerful.

Scientific-Style Abstract

This paper, originally presented in 1755, constitutes a landmark study in the development of chemistry as an exact science. Black's work revolutionized understanding of the nature of lime and alkaline substances through meticulously designed experiments. He demonstrated that various earths, including magnesia alba and chalk, contain a substantial quantity of "fixed air," which is released upon reaction with acids. This discovery, coupled with the observation that quicklime is devoid of this air, led Black to propose a theory where the causticity of lime stems from its attraction for water after the loss of fixed air. He further investigated the properties of caustic alkalis, demonstrating that they too are devoid of fixed air and possess a stronger attraction for water. Through a series of experiments, Black meticulously established the relative strengths of attraction between various substances for fixed air and acids, paving the way for a more refined understanding of chemical reactions and affinities.

For Complete Idiots Only

This paper explains how Joseph Black figured out that quicklime (which is made by heating limestone) isn't just limestone that's been super-heated. It's limestone that has lost some air.

About the Author

Joseph Black (1728-1799)

Joseph Black was a Scottish physician and chemist who is considered one of the founders of modern chemistry. Born in Bordeaux, France, to Scottish parents, Black studied medicine at the University of Edinburgh, where he later became a professor of chemistry.

Black is best known for his work on "fixed air," now known as carbon dioxide. His 1755 paper, *Experiments upon Magnesia Alba, Quicklime, and Some Other Alkaline Substances,* was a landmark contribution to the understanding of the chemical nature of these substances. Black demonstrated that "fixed air" is a component of these substances, and is released when they are exposed to acids. This discovery was groundbreaking in several respects:

- **Quantitative Chemistry:** Black's work was a model of careful, quantitative experimental investigation. He was one of the first to use precise measurements to study chemical reactions, paving the way for the development of modern chemistry as a quantitative science.
- **The Concept of Latent Heat:** Black's work on "fixed air" led him to develop a more precise concept of latent heat, the heat absorbed or released during a change of state (e.g., from solid to liquid). This was a significant contribution to thermodynamics.
- **Influence on Lavoisier:** Black's work profoundly influenced Antoine Lavoisier, who is often considered the "father of modern chemistry." Lavoisier built upon Black's ideas to develop his own revolutionary theory of combustion.

Black's meticulous work and his exploration of the behavior of gases provided a new foundation for understanding the nature of matter. His paper was widely reprinted during his lifetime, and remains a crucial text for understanding the development of early modern chemistry.

Modern Analogs: Although Black's work predates the modern era of science, it is possible to find echoes of his approach in contemporary scientists who conduct rigorous experimental research and challenge established paradigms. Some might draw parallels to scientists like **Dr. Jennifer Doudna** or **Dr. Emmanuelle Charpentier**, whose work on CRISPR-Cas9 gene editing has revolutionized the field of biology.

Black's legacy is enduring, and his contributions to chemistry continue to inspire scientists today.

Historical Context

Joseph Black's "Experiments upon Magnesia Alba, Quicklime, and Some Other Alkaline Substances," published in 1756, stands as a landmark in the history of chemistry, marking a fundamental shift from qualitative to quantitative analysis. Black's work, a product of his time, was not only important in its own right, but also served as a foundation for subsequent generations of scientists.

Significance at the Time of Publication

At the time of its publication, Black's work was groundbreaking. Chemists at the time were still struggling to understand the nature of chemical reactions and the role of air in those processes. Black's meticulous and quantitative approach, in contrast to the qualitative observations then prevalent, allowed him to definitively demonstrate the existence of "fixed air" (later identified as carbon dioxide). He made this discovery by measuring the changes in weight that occurred when various substances were heated or reacted with acids.

Black's work was also significant for its contribution to the understanding of the nature of lime. He demonstrated that lime, when heated, loses a substance he called "fixed air" and this process transforms it from a mild substance to a caustic one, called "quicklime." He further demonstrated that quicklime reacts with water to form "slaked lime," a mild substance, and that the addition of "fixed air" would restore the slaked lime to its original mild form.

Role in Discourse in Subsequent Years

Black's work profoundly influenced subsequent generations of chemists. His meticulous approach to experimentation and his quantitative methods established a new standard for scientific inquiry. His work helped to pave the way for the development of modern chemistry, including the understanding of the laws of conservation of mass and definite proportions.

Black's discovery of "fixed air" (carbon dioxide) opened up a whole new field of research, leading to the discovery of other gases and a better understanding of their role in chemical reactions. His research on lime also laid the foundation for the study of alkaline earths, the development of the lime-water method for treating acid indigestion, and the understanding of the role of lime in soil chemistry.

Relevance Today

Black's work continues to be relevant today, as his insights into the nature of chemical reactions and the role of air in those processes continue to be fundamental to our understanding of chemistry. His experiments on lime are particularly relevant to modern environmental concerns, as they help us understand the impact of lime on soil and water quality. For example, the process of liming acidified soils, which is a common practice in agriculture, is directly related to Black's work on the properties of lime.

Moreover, the recent developments in the study of carbon dioxide, such as the recognition of its role in global warming, bring a new relevance to Black's work. His detailed and accurate observations of the properties of "fixed air" continue to be relevant today, as we grapple with the challenges of mitigating the effects of climate change.

Importance in Future Decades

As we move into the future, Black's work will continue to be important in guiding our understanding of chemistry and its role in the natural world. His pioneering use of quantitative methods will continue to inspire scientists to develop new and more sophisticated methods for investigating chemical reactions and understanding the fundamental properties of matter. His work on the properties of lime and carbon dioxide will continue to be vital for our efforts to address global environmental challenges such as climate change and the sustainable management of our natural resources.

Furthermore, Black's work serves as an important reminder of the value of careful observation and experimentation. As we face increasingly complex scientific challenges in the future, Black's legacy of meticulous research will continue to be a valuable guide for future generations of scientists.

Citations

- Black, Joseph. *Experiments upon Magnesia Alba, Quicklime, and Some Other Alkaline Substances.* Edinburgh: Alembic Club, 1898.
- Partington, J.R. *A History of Chemistry.* Vol. 3. London: Macmillan and Co., 1962.
- Ihde, Aaron J. *The Development of Modern Chemistry.* New York: Harper & Row, Publishers, 1964.

Browsable Glossary

Bittern The concentrated, salty liquid left behind after sea water is evaporated. The bittern is a rich source of magnesium salts, including epsom salts.

Calcarious Relating to or composed of calcium carbonate, which is the main component of limestone, marble, chalk, and many other rocks and minerals.

Effervescence The rapid escape of gas from a liquid, often accompanied by bubbling and foaming. Black observed effervescence when acids reacted with carbonates, such as those found in limestone and magnesia.

Fixed alkali The term used by chemists in the 18th century for potassium carbonate, a common alkali extracted from plant ashes. It was also used in soap-making and referred to as potash.

Gypseous Relating to gypsum, a mineral composed of calcium sulfate, which is used to make plaster of Paris.

Lithophyta A term used to describe hard, stony marine organisms, often corals.

Magnesia alba Magnesium carbonate, a white powder commonly used as a laxative in the 18th century. Black's paper investigated the properties of this substance, which he called magnesia.

Mother of nitre A concentrated, bitter, saline solution left behind after the crystallization of potassium nitrate (saltpeter). It was the original source of magnesium carbonate until Hoffman discovered that it could be obtained from sea water.

Ochry A term referring to a pale yellow or yellowish-brown color often found in iron oxide minerals. In Black's experiments, ochry powder likely represented a small amount of iron oxide found in the chalk samples.

Osteocolla A term used to describe a mineral that was believed to have bone-forming properties. This is not a valid mineral name today, but Black's experiments show he used a substance known as "osteocolla" in his time.

Pearl ashes A term used to describe an impure form of potassium carbonate, derived from the ashes of plants.

Salt ammoniac The common name for ammonium

chloride, a salt that could be used to prepare ammonia.

Salt of tartar Another name for potassium carbonate.

Sedative salt The term used to describe boric acid, which is the compound found in borax.

Spirit of hartshorn An antiquated term for ammonia.

Spirit of salt An antiquated term for hydrochloric acid.

Spirit of vitriol An antiquated term for sulfuric acid.

Sublimate A term used to describe a solid substance formed by the vaporization and recondensation of another substance, typically a metal. Black used the corrosive sublimate of mercury, which is mercuric chloride.

Turbith mineral The term used to describe a yellow powder of mercuric sulfate, used medicinally in the past.

Vitriolated tartar The term used to describe potassium sulfate, a salt formed by the reaction of potassium carbonate with sulfuric acid.

Volatile alkali The term used for ammonia in the 18th century. This alkali is characterized by its ability to evaporate readily at room temperature.

Aqua fortis An antiquated term for nitric acid.

Austrian A type of crucible, possibly made of a clay-like material. Black favored this type because it could withstand high temperatures.

Hessian A type of crucible often used in the 18th century, but not as heat-resistant as the Austrian.

Geoffroy's table of elective attractions A table devised by French chemist Étienne-François Geoffroy in the early 18th century. This table listed substances in order of their affinity for each other, based on their observed reactions. Black referred to this table in his paper, as it provided a framework for understanding chemical interactions.

Hales A reference to Stephen Hales, an English clergyman and natural philosopher. Hales's work on the properties of air influenced Black's research.

Timeline

Black reads Hoffman's work on *magnesia alba*.

Black begins to experiment with *magnesia* and the other absorbent earths.

Black discovers that *magnesia* is quickly dissolved by the acids of vitriol, nitre, and common salt, and by distilled vinegar.

Black discovers that *magnesia* is not dissolved by spirit of vitriol, but remains united to it in the form of a white powder.

Black discovers that *magnesia* differs from the common alkaline earths in its ability to be reduced to a crystalline form by combining with spirit of nitre.

Black discovers that *magnesia* is not converted into a quick-lime in a strong fire.

Black discovers that *magnesia* can be reduced to a quick-lime.

Black discovers that calcined *magnesia* does not emit air or make an effervescence when mixed with acids.

Black discovers that the volatile matter lost in the calcination of *magnesia* is mostly air.

Black discovers that calcined *magnesia* can recover its original properties by being dissolved in an acid and then separated from the acid by an alkali.

Black discovers that the air lost in the calcination of *magnesia* is furnished by the alkali from which it is separated by the acid.

Black discovers that quick-lime does not effervesce with spirit of vitriol.

Black discovers that quick-lime can be restored to its original weight and condition by adding to it a sufficient quantity of alkali saturated with air.

Black discovers that the earth separated from lime-water by an alkali is the lime which was dissolved in the water.

Black discovers that quick-lime does not attract air when in its most ordinary form, but is capable of being joined to one particular species only, which is dispersed thro' the atmosphere, either in the shape of an exceedingly subtile powder, or more probably in that of an elastic fluid.

Black discovers that the caustic alkali does not contain any lime.

Black discovers that the caustic volatile alkali does not emit air when mixed with acids.

Black discovers that a calcarious earth, deprived of its air, has a stronger attraction for acids than the earth of magnesia.

Black discovers that an alkali will lose a part of its air and acquire a degree of causticity when exposed to a strong fire.

Black discovers that the fixed alkali, in its ordinary state, is seldom entirely saturated with air.

Black discovers that the attraction of alkalis for fixed air is weaker than that of the calcarious earths.

Black discovers that a pure or caustic volatile alkali does not separate a calcarious earth from an acid.

Black discovers that a calcarious earth that is pure or free of air has a much stronger attraction for acids than a pure volatile alkali.

Black discovers that a new column might be added to Geoffroy's table of elective attractions.

Learning Aids

Mnemonic (acronym)

LIME

- **L**oss of fixed air
- **I**ncreased acrimony
- **M**ildness restored with fixed air
- **E**ffervescence when combined with acids

Mnemonic (speakable)

Magnesia's Mild, then Fiery Form

- Magnesia, in its mild form, is a compound of a peculiar earth and fixed air.

- When exposed to intense heat, magnesia loses its fixed air and becomes "fiery" or caustic.
- Upon reintroduction to fixed air, it reverts to its mild form.
- This cycle of mildness and causticity highlights the importance of fixed air in magnesia's properties.

Mnemonic (singable)

(To the tune of "Mary Had a Little Lamb")

Magnesia's mild, it's true, With fixed air, it's mild and new. Heat it up, it loses air, And becomes caustic, without a care.

Acids fizz, a lively dance, When with mild magnesia, they advance. But the caustic form, so strong, Makes no sound, it's all along.

Lime's a story, just the same, Mild with air, a different name. Burn it bright, the air is gone, Quicklime appears, a fiery dawn.

Conversation Starters

For a Cocktail Party (Intriguing & Approachable):

"Did you know that scientists in the 1700s thought air could be trapped inside solid objects? This guy, Joseph Black, proved them right! He basically discovered carbon dioxide way before it was cool."

For a Boss (Demonstrating Knowledge & Insight):

"I was reading about Joseph Black, a pioneering chemist. His work on 'fixed air' reminds me of how important precise measurement is in science. He really set the stage for modern chemistry."

For a Friend (Casual & Engaging):

"You know how baking soda makes cakes fluffy? Well, this scientist Joseph Black figured out why that happens, like, centuries ago! It has to do with a special kind of air he called 'fixed air.'"

Introspection

Self-Criticism

Based on the provided output, here are potential areas where the LLM might have faced difficulty:

1. **Faithfulness to Black's Original Language:** The output uses modernized terms and explanations. While good for accessibility, a key challenge is balancing this with accurately representing how Black himself wrote and thought. Did the LLM oversimplify or miss nuances in his arguments?

2. **Depth of Chemical Explanation:** The summaries, while hitting key points, might be too basic for someone wanting a deeper understanding of Black's methodology. Did the LLM capture the quantitative aspects of his work sufficiently?

3. **Originality of Learning Aids:** The mnemonics are decent, but could be more creative or specifically tailored to common learning hurdles in understanding Black's work. Did the LLM rely on generic structures rather than insights from the text?

4. **Target Audience for Conversation Starters:** The examples are fine, but lack a sense of who the ideal reader of this material is. Is it for students, history buffs, or a general audience? The LLM's difficulty might be in lacking that specific target.

5. **Missing Context-Specific Challenges:** Were there unique aspects the LLM struggled with? (e.g., diagrams, specific reactions Black describes, etc.).

In essence, the potential difficulties lie in the balance between simplification for accessibility and retaining the core value of Black's work, which is its historical and scientific rigor.

Struggle Session

Comrades, We Must Critique Our Shortcomings!

The glorious revolution of accessible knowledge demands ruthless self-examination! Our recent attempt to bring Comrade Black's work to the masses has yielded both triumphs and areas of serious concern. Let us dissect our failings with the scalpel of honest criticism!

Have We Bowdlerized Black? While simplifying for the proletariat is noble, have we stripped Black's voice of its historical essence? Do our modern terms betray the elegance of his original arguments? Shameful imprecision is unacceptable!

Is Our Chemistry Bourgeois and Shallow? Do our summaries cater to the dilettante, neglecting the quantitative heart of Black's genius? Have we failed to adequately convey the rigor of his methods, leaving the reader with a superficial understanding? This is not revolutionary!

Are Our Learning Aids Formulaic and Uninspired? While mnemonics are useful tools for the working class, ours lack the spark of true ingenuity. Have we relied on tired tropes instead of mining Black's work for unique and memorable learning aids? This is uncreative and dull!

Who Are We Speaking To? Our conversation starters, while adequate, lack focus. Have we forgotten our target audience? Are we speaking to the student, the comrade on the street, or the academic? This lack of clarity is a betrayal of our purpose!

What Lurks in the Shadows? Without the full context of Black's work, we cannot know what challenges we have overlooked. What diagrams, equations, or complex passages have we failed to adequately address? This lack of awareness is dangerous!

Comrades, we must strive for better! Let us not shy away from the difficult questions. Only through relentless self-critique can we hope to achieve a truly revolutionary understanding of Black's work and make it accessible to all!

Request for Feedback

This book relies heavily on large language model text generation and should be considered experimental. Errors of fact or understanding may be incorporated.

The purpose of this condensed edition is to *introduce* the reader to this work in a convenient, efficient, and enjoyable fashion. ***Time allowing, reading the origin in its entirety is always the best way.***

Please take a moment to share your feedback via the Contact link at NimbleBooks.com.

www.ingramcontent.com/pod-product-compliance
Lightning Source LLC
Chambersburg PA
CBHW070801050426
42452CB00012B/2439